AF605479

Original Korean text by Mi-ae Lee
Illustrations by Dong-soo Kim

This English edition published by big & SMALL in 2015
by arrangement with Aram Publishing
English text edited by Joy Cowley

ISBN: 978-1-925233-74-2

Printed in Korea

Twinkle, Twinkle!

Written by Mi-ae Lee
Illustrated by Dong-soo Kim
Edited by Joy Cowley

It is evening, early summer
and the fireflies are out.
Twinkle, twinkle!
Small lights in their tails
dance like falling stars.

A female firefly sits
on a blade of grass.
Her light says she is lonely.

A male firefly sees her light.
He blinks his light at her,
and she answers.
Twinkle, twinkle.

It is raining.
The two fireflies hide
until the rain stops.
Then they come out to mate
on a blade of grass.

A few nights later,
the female firefly flies around
looking for a good place
to lay her eggs.

She finds a wet root covered with moss
near a stream of running water,
and she blinks her light
to let other female fireflies know
that this is the best place
to lay their little round eggs.

While the female firefly
is laying her eggs,
the male firefly blinks his light
for the last time.
He has done his work
and now his life is over.

When all her eggs are laid,
the female firefly also dies,
leaving new life on the wet moss.
Hundreds of tiny eggs
will become new fireflies.

After a while,
the soft transparent eggs
show signs of larvae.

The bodies of insects like fireflies change completely as they develop. Larvae come from the eggs then change into pupa, before finally emerging as adults. This process of changing form is called metamorphosis.

A spotted larva breaks the egg
with its sharp chin
and comes out into the world.

At once, it heads for the stream.

The larvae twist their bodies
to sink down, down, down,
to the bottom of the stream.
They crawl about looking for food.

The larvae breathe with gills
as they feed on black snails
and other small creatures.

When fish try to eat them,
they hide between rocks.

Two months have passed.
The larvae turn gray
and stay between stones
for about five hours.
Then they wriggle
and cast off their skins
for the first time.

Before spring comes,
the larvae will grow,
shedding their skins four times.
Then they swim to the surface.

A larva crawls into wet ground,
makes a nest and falls asleep.

When the larva wakes,
it casts off its skin again
and turns into a pupa.

The pupa casts off its skin for the last time.
Finally, it is a firefly.

The firefly's body must harden before it can come out of the ground.

A few nights later,
the firefly is out of the ground.

Although a lot of eggs are laid,
some are eaten by fish
and some by spiders and frogs.
But some eggs survive
and turn into fireflies.

In early summer,
the survivors fly high,
blinking their tail lights
as they look for a mate.

Twinkle! Twinkle!

Twinkle, Twinkle!

Fireflies are a good example of how most insects live. Insects have four stages in their life cycle: egg, larva, pupa, and adult. Adults reproduce to start the life cycle again. Although insects have various characteristics and habits, most follow this life cycle. Some insects, like grasshoppers, have different life stages.

Let's think

Why do insects lay lots of eggs?

What is metamorphosis?

Why do adult insects die soon after reproducing?

Where are you likely to find insects at the different stages of their life cycle?

Let's Do!

It's time for a field study. Let's go and explore, but don't forget to take your notepad and a pencil! Find an outdoor area with lots of trees and bushes, and a pond or a stream if possible. Look for insects: adults, eggs, larvae, pupa or other forms. A pond or a stream is a good place for finding larvae. Draw pictures and write down what you can find.